I0797688

PRAIRIE DOGS

by Martha London

Cody Koala
An Imprint of Pop!
popbooksonline.com

abdobooks.com
Published by Pop!, a division of ABDO, PO Box 398166, Minneapolis, Minnesota 55439.

Printed in the United States of America, North Mankato, Minnesota

082020
012021

Cover Photo: Shutterstock Images
Interior Photos: Shutterstock Images, 1, 5 (top), 5 (bottom right), 10–11, 13, 15 (top); iStockphoto, 5 (bottom left), 6, 15 (bottom right), 15 (bottom left), 19, 20; David Schleser/Nature's Images/Science Source, 9; Tom & Pat Leeson/Science Source, 16–17

Editors: Christine Ha and Brienna Rossiter
Series Designer: Sophie Geister-Jones

Library of Congress Control Number: 2019955011

Publisher's Cataloging-in-Publication Data
Names: London, Martha, author.
Title: Prairie dogs / by Martha London
Description: Minneapolis, Minnesota : POP!, 2021 | Series: Underground animals | Includes online resources and index.
Identifiers: ISBN 9781532167645 (lib. bdg.) | ISBN 9781532168741 (ebook)
Subjects: LCSH: Prairie dogs--Juvenile literature. | Rodents--Juvenile literature. | Ground squirrels--Juvenile literature. | Burrowing animals--Juvenile literature. | Underground areas--Juvenile literature.
Classification: DDC 599.367--dc23

Hello! My name is

Cody Koala

Pop open this book and you'll find QR codes like this one, loaded with information, so you can learn even more!

Scan this code* and others like it while you read, or visit the website below to make this book pop.

popbooksonline.com/prairie-dogs

*Scanning QR codes requires a web-enabled smart device with a QR code reader app and a camera.

Table of Contents

Chapter 1

A Big Burrow

Prairie dogs are **mammals**. They live in large groups called **colonies**. Prairie dogs dig homes known as **burrows**. Each burrow has many **entrances**.

Watch a video here!

Prairie dogs live in North America. They dig their burrows in flat, grassy places. Prairie dogs pile dirt by each entrance. They stand on these mounds to watch for **predators**. If the prairie dogs sense danger, they hide in the burrow.

Each burrow has many tunnels and rooms. Some rooms are for sleeping. Others are for storing food. There are even rooms for going to the bathroom.

A prairie dog burrow can have 60 entrances.

Chapter 2

Staying Safe

Prairie dogs have tan fur. Their fur blends in with dirt and grass. It makes prairie dogs hard for **predators** to see.

Learn more here!

Prairie dogs have sharp claws. They use their claws to dig **burrows**. Prairie dogs also have good hearing and eyesight. They use these senses to look out for danger.

Prairie dogs are approximately 12 inches (30 cm) tall.

eye
ear
mouth
claws
tail

Chapter 3

Working Together

Prairie dogs eat seeds, grasses, and roots. They leave their **burrows** to look for food. Then the prairie dogs in a **colony** work together.

Learn more here!

Some prairie dogs collect food. Others stay close to the burrow. These prairie dogs are lookouts. They take turns watching for **predators**.

If a predator is near, the lookouts bark to **alert** the others. All the prairie dogs hide. They stay hidden until the predator leaves.

Chapter 4

Large Families

Several prairie dog families live together in a **colony**. Each family is called a **coterie**. It includes one male, a few females, and their babies.

Complete an activity here!

Baby prairie dogs are called pups. The females in a colony work together to raise them. Most pups stay in the colony when they grow up.

Prairie dogs live for approximately three to five years in the wild.

Making Connections

Text-to-Self

Would you want to live near a prairie dog colony? Why or why not?

Text-to-Text

What books have you read about other animals that live underground? What kind of homes did those animals make?

Text-to-World

Many prairie dogs live together in a colony. What other animals live in large groups?

Glossary

alert – to warn others of danger.

burrow – a hole that an animal digs in the ground for shelter.

colony – a group of similar animals that live together in one place.

coterie – a family of prairie dogs.

entrance – a way to get inside a place.

mammal – a type of animal that has hair or fur and feeds milk to its young.

predator – an animal that hunts other animals for food.

Index

Online Resources

popbooksonline.com

Thanks for reading this Cody Koala book!

Scan this code* and others like it in this book, or visit the website below to make this book pop!

popbooksonline.com/prairie-dogs

*Scanning QR codes requires a web-enabled smart device with a QR code reader app and a camera.